TIMES TABLES THE FUN WAY™

STUDENT WORKBOOK

Second Edition Revised

© 1995

Key Publishers, Inc.
Sandy, Utah

ISBN 1-883841-33-X

Written by: Judy Rodriguez

Illustrated by:
Val Bagley and Judy Rodriguez

TIMES TABLES THE FUN WAY™

STUDENT WORKBOOK
Second Edition Revised

For Information:
KEY PUBLISHERS, INC.,
6 SUNWOOD LANE
SANDY, UTAH 84092
1-801-572-1000

Copyright © 1993, 1995 by Key Publishers, Inc.
First Printing: 1993
Second Printing: 1994
Third Printing, Revised: 1995
Fourth Printing:1996
ISBN 1-883841-33-X

Attention: Schools and Businesses
Times Tables the Fun Way Student Workbook is available at quantity discounts with bulk purchase for educational, business, or sales promotional use.

NOTE TO TEACHERS AND PARENTS

Times Tables the Fun Way is a picture and story method of learning the times tables. Students enjoy the variety and range of activities that are possible with the story method. This workbook should be used in conjunction with the *Times Tables the Fun Way* Book for Kids, (Student Text), and *Times Tables the Fun Way* Flash Cards. The *Times Tables the Fun Way* Teacher's Manual is designed for classroom or workshop use. The Teacher's Manual contains detailed lesson plans, games, game cards, and grading templates. Parents or teachers working one-on-one can teach the *Times Tables the Fun Way* method with the student text, workbook, and flash cards, (Teacher's Manual optional) by following the Master Plan included in this workbook.

The success of this method depends on adequate practice in applying the stories to the raw fact. Studies have shown that students who learn with the picture story method retain the facts for longer periods and score higher on post-tests because the facts are firmly implanted in long term memory. In the early learning phases, it will be necessary to give story clues to students in order to trigger the answers to the facts. Once the stories are thoroughly learned and practiced with the workbook and text, students will by-pass the story when remembering the fact. So, the stories are an intermittent, but crucial phase of learning the multiplication facts by the picture and story method.

Be sure to use the workbook pages in the order presented and follow the Master Plan. Otherwise, students will encounter facts that have not yet been taught. While teaching each lesson, always give the quiz first. Next, read, teach, and study the text book pages. Then do the workbook pages. The *Times Tables the Fun Way* system is designed as an introduction to times tables for 2nd or 3rd graders, but it can also be used in special education programs and in the higher grades for review or learning of the more difficult or forgotten facts.

Have fun teaching. We are sure that you and your students will enjoy learning *Times Tables the Fun Way*.

DESCRIPTION OF *TIMES TABLES THE FUN WAY*
WORKBOOK MATERIALS

Student Questionnaire:

The questionnaire is used to stimulate discussion. It starts the program with a personal and fun activity.

Pre-test:

The pre-test is timed for 6 minutes and used as a baseline to measure progress against the post-test. It contains 48 facts, some zeroes and all of the ones through nines, but no reversals, i.e. 7 x 6 but not 6 x 7. The commutative property is explained in the *Times Tables the Fun Way* Text.

Quizzes:

Quizzes are given at the start of each lesson. They are used to reinforce the stories by asking the students to draw a picture or write the story. This strengthens the connection between the story and the fact. It is always better to review the stories or give hints during the quiz, if necessary, than to leave the quiz section blank. Students learn by re-telling or drawing the picture of the story. The quiz has two parts. The upper portion is a test of the raw facts and the lower portion tests the student's memory of the stories. The story section should be marked correct if the students show any indication that they have remembered the correct story for that fact.

Timed Practice Sheets:

These sheets are timed and used to practice the ones, twos, fives, and nines. Students should write down the time it took them to complete the page. The goal is for students to improve their own times and scores on each type of Timed Practice Sheet. For example: The Ones and Twos Practice Sheet should get faster each time it is taken. However, the Fives Practice Sheet time should not be faster than the Ones and Twos because the Fives are harder. The Fives Practice Sheet Time should be faster than the last time the student took the Fives Practice Sheet.

Worksheets:

These are used to practice the introduced facts. Each worksheet contains only the facts that have been introduced so far. Students will learn to do two digit by one digit multiplication. At this stage, learning to add and multiply in the same problem adds diversity to times tables practice. The workbook pages go through a step by step explanation of two digit by one digit multiplication as well as simple division and learning the nines.

Post-test:

The post-test is timed for 6 minutes. The goal is for the students to score 100% on the post-test. The average score at the end of the 8 hour program in a workshop setting has been 97%.

Other Activity Sheets:

Homework Test: Students should take the graded test and learn the missed facts for homework. The Homework Quiz is given at the next lesson and will test the students on the facts that were missed on the Homework Test.

Crossword Puzzle: Facts are given as clues and students must fill in the key words of the story. This can be played in a group format or done individually.

DESCRIPTION OF WORKBOOK MATERIALS CONTINUED:

Story Quiz: Students list one or two key words of each story. If they miss any, they can go back and look up the story and fill in the correct key words.

Other Activity Sheets: pages with an A or B next to the page number are optional. Directions are outlined on each sheet. These educational activity sheets are new to the second edition.

Stamp and Score Summary Sheet:

Students are able to map their progress by filling out their Quiz scores and Timed Practice scores and times. There is a square at the bottom of the sheet to put a stamp or a sticker when the indicated goals are reached. Students should be rewarded after each lesson for improving scores, receiving 100%, winning at the games, or completing worksheet pages.

SKILLS COVERED IN THE STUDENT WORKBOOK
Multiplication Facts: 0 - 12
Double Digit By Single Digit Multiplication
Double Digit By Single Digit Multiplication With Regrouping
Simple Division

Memorizing the facts and rehearsing them over and over can be boring and tedious for students. When new, but simple skills are introduced during the facts learning process, students are able to apply their newly learned facts to varied operations. This teaches students how they will be using their multiplication facts while reinforcing the newly learned facts. When students encounter these skills in their grade level text book, they will already have been introduced to them, making the learning process easier and more enjoyable.

MEMORIZING THE TENS, ELEVENS, AND TWELVES

The *Times Tables The Fun Way* Method teaches students to memorize the facts, ones through the nines. Students are taught double digit multiplication in the second lesson and can easily figure out the answers for the 10's, 11's, and 12's. It is not necessary to have these facts memorized in order to do higher grade level skills like fractions, division, and decimals. Therefore, we do not feel that it is necessary to memorize these double digit facts.

DESCRIPTION OF *TIMES TABLES THE FUN WAY*
BOOK FOR KIDS

Times Tables the Fun Way Book is in full color and printed on heavy, durable pages. The text covers the zeroes through the nines. All facts for the threes, fours, sixes, sevens, and eights are taught with pictures and stories. Numerical tricks are used to teach the zeroes, ones, twos, fives, and nines. Each fact is addressed once and not taught in reverse order. The commutative property is explained so students learn, for example, that 6 x 7 is the same as 7 x 6.

DESCRIPTION OF *TIMES TABLES THE FUN WAY*
FLASH CARDS

The flash cards are an integral part of the picture story method. The cards for the story facts show the number characters as well as the raw numbers. The characters with the numbers help the students transition from the story to the fact. The number at the bottom right corner of the card represents the lesson that the fact is introduced. The cards can be sorted and used in the order of presentation.

MASTER LESSON PLAN

Lesson One

Facts Taught:
3 x 3 - 4 x 3 - 1's & 2's
8 x 8 - 7 x 7

Timed Practice:
1's & 2's

Pre-test

Book For Kids Pages:
10-32, 68-69, 58-59

Workbook Pages:
1-3

Lesson Two

Facts Taught:
4 x 4 - 8 x 7 - 6 x 6
5's - 6 x 4

Timed Practice:
5's

Quiz: #1

Book For Kids Pages:
33-40, 44-45, 66-67, 46-47

Workbook Pages:
4-14

Lesson Three

Facts Taught:
6 x 3 - 6 x 8

Timed Practice:
1's & 2's - 9's

Quiz: #2

Book For Kids Pages:
42-43, 48-49, 50, 71-81

Workbook Pages:
15-19

Lesson Four

Facts Taught:
3 x 7 - 7 x 4

Timed Practice:
5's

Quiz: #3

Book For Kids Pages:
52-53, 54-55

Workbook Pages:
20-26

Lesson Five

Facts Taught:
7 x 6 - 8 x 4

Timed Practice:
9's

Quiz: #4

Book For Kids Pages:
56-57, 65-66

Workbook Pages:
27-32

Lesson Six

Facts Taught:
8 x 3

Timed Practice:
5's

Quiz: #5

Book For Kids Pages:
62-63, 70, 82

Workbook Pages:
33-39

Lesson Seven

Facts Reviewed

Timed Practice:
9's

Quiz: #6

Workbook Pages:
40-42

Lesson Eight

Facts Reviewed

Post-test

Workbook Pages:
43-44

"TIMES TABLES THE FUN WAY"
WORKBOOK

STUDENT QUESTIONNAIRE

1. What is your full name?_____

2. Do you have a nick name? _____

3. What do you like best about school?

_____

4. What is your favorite subject? _____

5. What is your favorite food? _____

6. Do you have a pet, if so, what kind?

_____

7. If you could have one wish, any wish, what would it be?

_____

8. What are your favorite things to do? _____

_____

9. What do you want to be when you grow up?

_____

10. Name one thing that you are very proud of:

_____

"TIMES TABLES THE FUN WAY"
WORKBOOK Lesson 1
PRE-TEST

# correct:	% score:
$\overline{48}$	

NAME_____DATE_____TIME_____

2 x1	3 x2	4 x4	3 x6	7 x8	8 x9	1 x3	2 x4
5 x3	6 x4	8 x8	9 x4	1 x1	2 x2	5 x4	6 x6
9 x3	6 x9	7 x4	2 x9	4 x1	5 x2	5 x5	8 x6
9 x7	5 x7	9 x1	1 x5	0 x7	3 x7	6 x1	6 x7
8 x5	9 x9	3 x8	5 x9	4 x3	8 x1	7 x2	3 x0
7 x7	1 x7	2 x6	8 x4	5 x6	0 x9	3 x3	2 x8

"TIMES TABLES THE FUN WAY"
WORKBOOK Lesson 1
ONES AND TWOS TIMED PRACTICE

# correct:	% score:
___ 20	

1 x 1	1 x 2	3 x 1	6 x 1
8 x 1	9 x 1	7 x 1	5 x 1
1 x 4	2 x 2	5 x 2	2 x 6
2 x 9	8 x 2	7 x 2	2 x 4
2 x 3	3 x 3	4 x 3	497 x1

"TIMES TABLES THE FUN WAY"

GUESS THE FACT

Can you figure this one out?

Directions:
1. Write the fact in the blanks below.
2. Write the answer to the fact.
3. Color the pictures.

_____ = _____
Fact .... Answer

_____ = _____
Fact .... Answer

_____ = _____
Fact .... Answer

_____ = _____
Fact .... Answer

"TIMES TABLES THE FUN WAY"
HELP THE PIRATE
FIND THE TREASURE

<u>Step One</u>: Fill in the answers to these facts:

2	4	7	2	5	2	2	3
x 6	x 2	x 2	x 8	x 2	x 3	x 9	x 1
___	___	___	___	___	___	___	___
Clue #1	Clue #2	Clue #3	Clue #4	Clue #5	Clue #6	Clue #7	Clue #8

<u>Step Two</u>: Match the answer for each clue with the picture and write the clue word in the blanks below.

14 - sharks

8 - dolphin

18 -treasure map

12 - Africa

10 - wise owl

6 - mountain top

3 - treasure

16 - donkey

1. The treasure is in _____.

2. Ride the _____to cross the bay.

3. Watch out for_____ along the way.

4. Take a _____to the mountains.

5. Talk to the _____.

6. Go to the _____.

7. Find the _____in the tree.

8.Great! You helped the pirate find the hidden_____.

"TIMES TABLES THE FUN WAY"
WORKBOOK Lesson 2

QUIZ # 1   NAME_____DATE_____

Answer these facts:

3	7	3	8
x 3	x 7	x 4	x 8

CORRECT
_____
4

What is the story for 8 x 8 ?
Draw it or write it.

CORRECT
_____
1

What is the story for 3 x 3 ?
Draw it or write it.

CORRECT
_____
1

What is the story for 7 x 7 ?
Draw it or write it.

CORRECT
_____
1

What is the story for 3 x 4 ?
Draw it or write it.

CORRECT
_____
1

"TIMES TABLES THE FUN WAY"
WORKBOOK Lesson 2
FIVES TIMED PRACTICE

# correct:	% score:
$\dfrac{}{20}$	

NAME_____DATE_____TIME_____

1 x 5	5 x 5	9 x 5	5 x 4
6 x 5	4 x 5	8 x 5	5 x 3
7 x 5	3 x 5	5 x 2	5 x 8
5 x 1	5 x 9	5 x 6	5 x 7
3 x 3	4 x 3	8 x 8	7 x 7

"TIMES TABLES THE FUN WAY"
WORKBOOK Lesson 2
LEARNING DOUBLE DIGIT MULTIPLICATION

Let's do one together , and then you'll be able to do it yourself.

First, ask yourself what is 4 x 2?

$$\begin{array}{r} 5\,4 \\ \times\,2 \\ \hline ? \end{array}$$

Right! the answer is 8.
   So, put the 8 right under the 2.

$$\begin{array}{r} 5\,4 \\ \times\,2 \\ \hline 8 \end{array}$$

Now ask yourself, what is 5 x 2?

$$\begin{array}{r} 5\,4 \\ \times\,2 \\ \hline ? \end{array}$$

Right, the answer is 10, so put the 10 with it's 0 right under the 5.

$$\begin{array}{r} 5\,4 \\ \times\,2 \\ \hline 1\,0\,8 \end{array}$$

The most important thing to learn is to keep your numbers lined up in very neat columns. This is important because it will help you get the right answer when you learn to do problems like:

$$\begin{array}{r} 54 \\ \times\,54 \\ \hline \end{array}$$

Go on to the next page

"TIMES TABLES THE FUN WAY"
WORKBOOK Lesson 2
LEARNING DOUBLE DIGIT *Page Two*

Now try this one yourself:

Ask yourself what is 2 x 3? Put your answer in the box:

$$\begin{array}{r} 6\,3 \\ \underline{x\,2} \\ \end{array}$$

Now ask yourself what is 6 x 2? Put your answer in the boxes to the left of your 6:

$$\begin{array}{r} 6\,3 \\ \underline{x\,2} \\ 6 \end{array}$$

Good Job ! Now try these:

$$\begin{array}{r} 3\,5 \\ \underline{x\,1} \end{array} \qquad \begin{array}{r} 2\,3 \\ \underline{x\,2} \end{array} \qquad \begin{array}{r} 4\,2 \\ \underline{x\,3} \end{array} \qquad \begin{array}{r} 3\,4 \\ \underline{x\,2} \end{array} \qquad \begin{array}{r} 2\,4 \\ \underline{x\,2} \end{array}$$

You're doing great! Can you do these? Be sure to keep your columns straight.

$$\begin{array}{r} 29 \\ \underline{x1} \end{array} \qquad \begin{array}{r} 31 \\ \underline{x3} \end{array} \qquad \begin{array}{r} 32 \\ \underline{x3} \end{array} \qquad \begin{array}{r} 41 \\ \underline{x4} \end{array} \qquad \begin{array}{r} 43 \\ \underline{x2} \end{array}$$

Let's do one together , and then you'll be able to do it yourself.

First, ask yourself what is 4 x 3?

```
  5 4
x   3
    ?
```

Right! The answer is 12.
But there is only one column for the 12, so you have to squeeze the 1 into its waiting place......

```
  1
  5 4
x   3
    2
```

Now ask yourself, what is 5 x 3?

```
  5 4
x   3
  ?
```

Right, the answer is 15, but before you write it you have to add the number in the waiting place, so ask yourself what is 15 + 1 ?

Right, 15 + 1 is 16. Now you can write the answer in its column.

```
  5 4
x   3
1 6 2
```

Go on to the next page

Now try this one:

- -

Ask yourself what is 4 x 3? Write your answer:

$$
\begin{array}{r}
6\,3 \\
x\,4 \\
\hline
\end{array}
$$

- -

Now ask yourself what is 6 x 4?

$$
\begin{array}{r}
6\,3 \\
x\,4 \\
\end{array}
$$
?

- -

Ok, now keep the answer to 6 x 4 in your mind while you add the number from the waiting place.

Now write your answer.

$$
\begin{array}{r}
6\,3 \\
x\,4 \\
\hline
\end{array}
$$

- -

Good Job ! Now try these:

$$
\begin{array}{r}
2\,5 \\
x\,2 \\
\hline
\end{array}
\qquad
\begin{array}{r}
3\,6 \\
x\,2 \\
\hline
\end{array}
\qquad
\begin{array}{r}
2\,8 \\
x\,2 \\
\hline
\end{array}
\qquad
\begin{array}{r}
2\,2 \\
x\,7 \\
\hline
\end{array}
\qquad
\begin{array}{r}
5\,3 \\
x\,4 \\
\hline
\end{array}
$$

Go on to the next page

"TIMES TABLES THE FUN WAY"
WORKBOOK Lesson 2

DOUBLE DIGIT WITH REGROUPING *Page Three*

Excellent! Now try these: Be sure to keep your columns straight.

$$
\begin{array}{r} 6\,2 \\ \times\,6 \\ \hline \end{array}
\qquad
\begin{array}{r} 8\,2 \\ \times\,8 \\ \hline \end{array}
\qquad
\begin{array}{r} 1\,8 \\ \times\,7 \\ \hline \end{array}
\qquad
\begin{array}{r} 2\,7 \\ \times\,7 \\ \hline \end{array}
\qquad
\begin{array}{r} 5\,5 \\ \times\,5 \\ \hline \end{array}
$$

Now check your answers for pages 7, 9, and 10 by turning to the back of the book. Then put your number correct here:

correct
___
22

Let's try doing some real easy stuff. Did you remember that zero times any number is always zero, because zero is the chief?

See if you can do these:

$$
\begin{array}{r} 0 \\ \times 1 \\ \hline \end{array}
\ \begin{array}{r} 7 \\ \times 0 \\ \hline \end{array}
\ \begin{array}{r} 0 \\ \times 8 \\ \hline \end{array}
\ \begin{array}{r} 9 \\ \times 0 \\ \hline \end{array}
\ \begin{array}{r} 6 \\ \times 0 \\ \hline \end{array}
\ \begin{array}{r} 0 \\ \times 3 \\ \hline \end{array}
\ \begin{array}{r} 4 \\ \times 0 \\ \hline \end{array}
\ \begin{array}{r} 5 \\ \times 0 \\ \hline \end{array}
\ \begin{array}{r} 8 \\ \times 0 \\ \hline \end{array}
\ \begin{array}{r} 95 \\ \times 0 \\ \hline \end{array}
$$

Wow! That was easy!

"TIMES TABLES THE FUN WAY"
WORKBOOK Lesson 2
TENS ELEVENS AND TWELVES

REMEMBER THAT ZERO TIMES ANY NUMBER IS ZERO.

11	10	12	10	11	12	10
x2	x3	x3	x2	x1	x2	x5

12	10	12	11	11	10	12
x5	x7	x6	x7	x9	x9	x4

12	10	11	10	11	12	10
x1	x8	x8	x1	x6	x9	x4

12	11	11	10	11	10	12
x8	x3	x4	x6	x5	x5	x2

Congratulations! Now you know how to figure out the tens, elevens, and twelves.

Go on to the next page

12	11	14	33	17
x3	x4	x3	x3	x7

18	17	22	33	32
x8	x2	x7	x4	x3

36	24	11	12	22
x2	x3	x9	x9	x8

27	28	12	12	12
x7	x8	x7	x5	x4

"TIMES TABLES THE FUN WAY"
WORKBOOK Lesson 2
WORKSHEET # 1a

25 <u>x3</u>	45 <u>x4</u>	43 <u>x3</u>	52 <u>x3</u>	57 <u>x7</u>
28 <u>x8</u>	47 <u>x2</u>	22 <u>x5</u>	32 <u>x4</u>	55 <u>x3</u>
78 <u>x2</u>	44 <u>x3</u>	21 <u>x9</u>	66 <u>x6</u>	77 <u>x8</u>
28 <u>x7</u>	27 <u>x8</u>	72 <u>x7</u>	52 <u>x5</u>	32 <u>x4</u>

52	44	26	52	27
x3	x5	x6	x5	x7
18	58	66	32	35
x8	x2	x5	x2	x5
65	33	23	61	88
x2	x4	x3	x6	x7
77	28	82	72	32
x7	x8	x7	x5	x3

"TIMES TABLES THE FUN WAY"
GUESS THE FACT

Gee, I love this book!

Times Tables The Fun Way

Directions:
1. Write the fact in the blanks below.
2. Write the answer to the fact.
3. Color the pictures.

= _____
Fact _____ Answer

= _____
Fact _____ Answer

= _____
Fact _____ Answer

= _____
Fact _____ Answer

"TIMES TABLES THE FUN WAY"

Directions:
1. Find the answer for one problem on Dufus.
2. Use the color key to find the color for that section.
3. Color the section.
4. Repeat until Dufus is competely colored.

Color Key
5 = White
10 = Brown
15 = Red
20 = Pink
30 = Black
35 = Yellow
40 = Green
45 = Blue

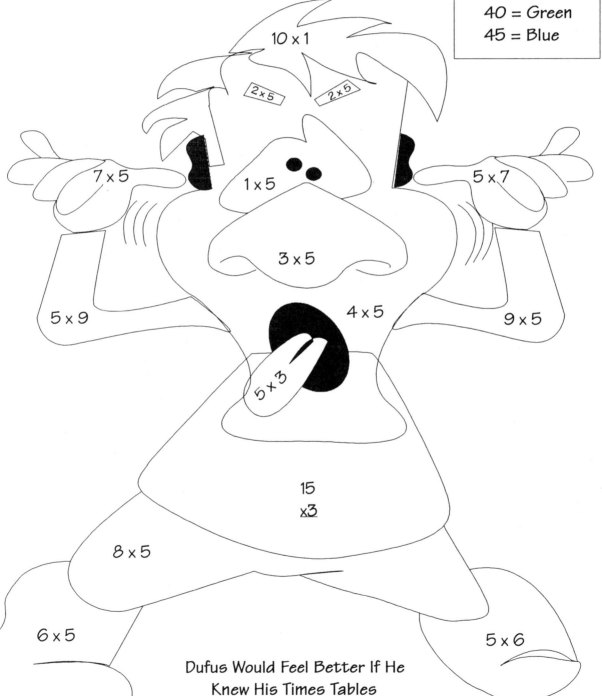

Dufus Would Feel Better If He
Knew His Times Tables

"TIMES TABLES THE FUN WAY"
WORKBOOK Lesson 3

QUIZ # 2 NAME_____DATE_____

Answer these facts:

6	4	8	6	8	7	3	4
x 4	x 4	x 7	x 6	x 8	x 7	x 3	x3

CORRECT
___
8

What is the story for 6 x 4 ?
Draw it or write it.

CORRECT
___
1

What is the story for 8 x 7 ?
Draw it or write it.

CORRECT
___
1

What is the story for 4 x 4 ?
Draw it or write it.

CORRECT
___
1

What is the story for 6 x 6 ?
Draw it or write it.

CORRECT
___
1

"TIMES TABLES THE FUN WAY"
WORKBOOK Lesson 3
ONES AND TWOS TIMED PRACTICE

# correct:	% score:
$\dfrac{}{20}$	

NAME_____ DATE_____ TIME_____

1 x 1	1 x 2	3 x 1	6 x 1
8 x 1	9 x 1	7 x 1	5 x 1
1 x 4	2 x 2	5 x 2	2 x 6
2 x 9	8 x 2	7 x 2	2 x 4
2 x 3	3 x 3	4 x 3	497 x 1

LEARNING NINES

First, write down the number that is one less. Sample: What is one less than 5 ? Right! It's 4, so put a 4 in the blank | 5 _4_ |

3___   7___   8___   6___   4___   5___   9___   2___

Now, write down the number that you would add to make nine. Sample: | 3 _6_ | **Because 3 + *6* = 9.**

2___   4___   5___   6___   9___   7___   8___   3___

On the next page, we will put these two steps together like this:

1. In the first space put the number that is one less than the number that nine is multiplied by:	2. In the space with the double line put the number that you would add to make nine (2 + 7 = 9, 3 + 6 = 9):
Sample:	**Sample:**
9 4	**9** 4
x3 **x9**	**x3** **x9**
_2_ _3_	_2 7_ _3 6_

Go on to the next page

"TIMES TABLES THE FUN WAY"
WORKBOOK Lesson 3
LEARNING NINES *Page Two*

Be careful!! Sometimes the number that nine is multiplied by is on top and sometimes its on the bottom. Always use the number that is not the nine to figure out one less. (Except, of course, when it's 9 x 9.)

9	9	③	9	1	8	7
x7	x⑥	**x9**	x2	**x9**	**x9**	**x9**

one less   one less

6 3	⑤ 4	② 7	— —	— —	— —	— —
	5 + 4 = 9	2 + 7 = 9				

9	9	6	9	7	9	9
x8	x1	**x9**	**x9**	**x9**	x5	x4

— —	— —	— —	— —	— —	— —	— —

9	9	3	2	5	9	9
x3	x2	**x9**	**x9**	**x9**	x6	x9

— —	— —	— —	— —	— —	— —	— —

9	8	9	9	5	9	9
x7	**x9**	**x2**	**x4**	**x9**	**x9**	**x8**

— —	— —	— —	— —	— —	— —	— —

Congratulations! Now you know your nines!

"TIMES TABLES THE FUN WAY"
WORKBOOK Lesson 3
NINES TIMED PRACTICE

# correct:	% score:
___ 20	

NAME_____DATE_____TIME_____

9 x 5	8 x 9	4 x 9	9 x 9
3 x 9	6 x9	9 x 7	9 x 1
2 x9	9 x 4	5 x 9	9 x 2
9 x 8	9 x 6	7 x 9	9 x 3
7 x7	4 x 3	8 x 8	6 x 4

"TIMES TABLES THE FUN WAY"

Directions:
1. Find the answer for one problem on the Math Wizard.
2. Use the color key to find the color for that section.
3. Color the section.
4. Repeat until the Math Wizard is competely colored.

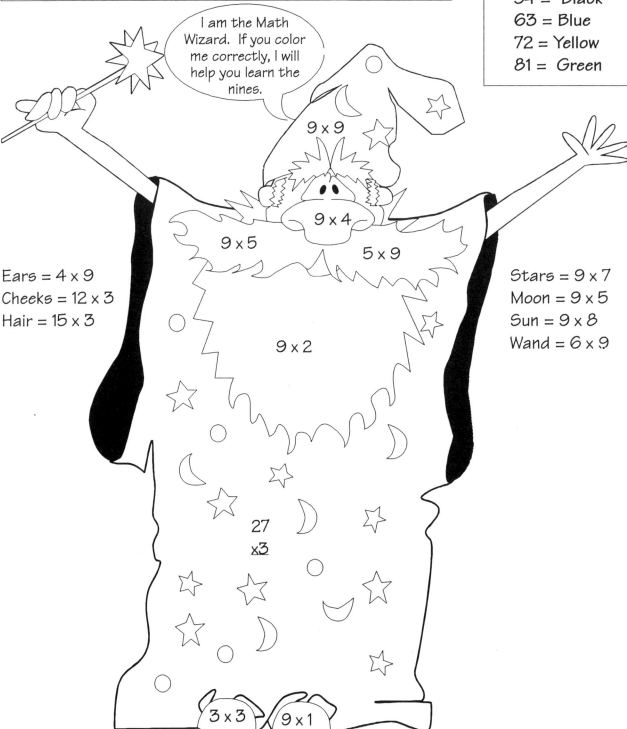

"TIMES TABLES THE FUN WAY"
CONNECT THE DOTS

Directions:
1. Start at the ☒ Count by 5's to connect the dots.
2. When you reach 60 draw your line to the 5 and continue.

Use me to connect the dots.

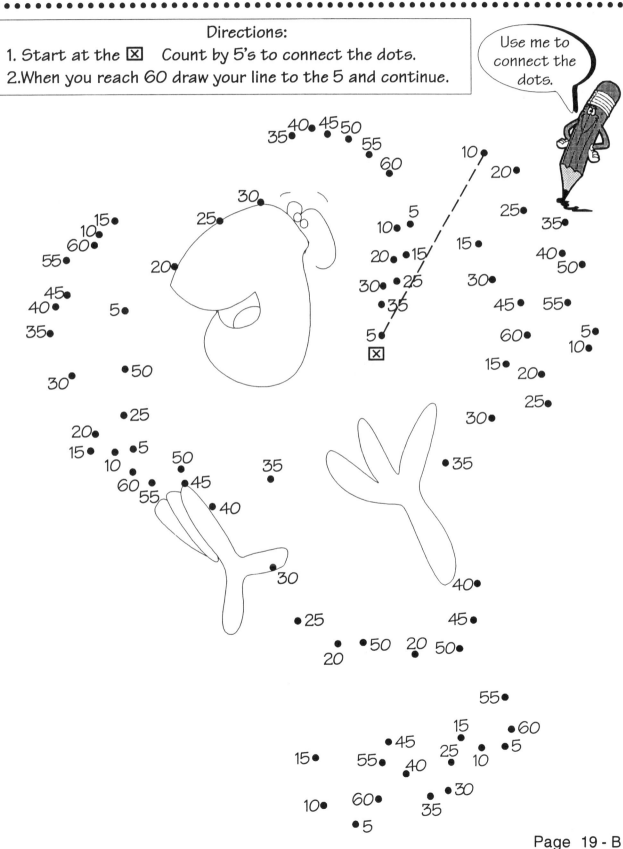

"TIMES TABLES THE FUN WAY"
WORKBOOK Lesson 4

QUIZ # 3 NAME_____DATE_____

Answer these facts:

6	6	5	6	9	7	8	4
x 3	x 8	x 5	x 5	x 9	x 9	x 5	x9

CORRECT
_____
8

What is the story for 6 x 3 ? Draw it or write it.

CORRECT
_____
1

What is the story for 6 x 8 ? Draw it or write it.

CORRECT
_____
1

"TIMES TABLES THE FUN WAY"
WORKBOOK Lesson 4
FIVES TIMED PRACTICE

# correct:	% score:
___ 20	

NAME_____DATE_____TIME_____

1 x 5	5 x 5	9 x 5	5 x 4
6 x 5	4 x 5	8 x 5	5 x 3
7 x 5	3 x 5	5 x 2	5 x 8
5 x 1	5 x 9	5 x 6	5 x 7
3 x 3	4 x 3	8 x 8	7 x 7

"TIMES TABLES THE FUN WAY"
WORKBOOK Lesson 4
LEARNING DIVISION

Division is the opposite of multiplication. Here is a sample: $64 \div 8 = ?$

Ask yourself what number times 8 will give you 64. Right! The answer is 8 because $8 \times 8 = 64$.

Here's another one: $56 \div 7 = ?$
What number times 7 will give you 56?
Right! The answer is 8 because $7 \times 8 = 56$

Try these:

$25 \div 5 =$ _____ (*Think:* **25 = 5 x ?**)
$12 \div 3 =$ _____ (*Think:* **12 = 3 x ?**)
$16 \div 4 =$ _____ (*Think:* **16 = 4 x ?**)
$18 \div 3 =$ _____ (*Think:* **18 = 3 x ?**)
$30 \div 6 =$ _____ (*Think:* **30 = 6 x ?**)
$49 \div 7 =$ _____ (*Think:* **49 = 7 x ?**)
$81 \div 9 =$ _____ (*Think:* **81 = 9 x ?**)
$40 \div 5 =$ _____ (*Think:* **40 = 5 x ?**)

Go on to the next page

There is another way to write division problems. They look like this:

$$6\overline{)36}$$

When you see a problem like this, you ask yourself: What number times 6 equals 36, or 6 x ? = 36. The answer is 6 because 6 x 6 = 36. So, put the 6 on top of the house.

$$6\overline{)36}$$ with 6 on top

Now try these:

$$4\overline{)16} \quad 3\overline{)12} \quad 5\overline{)40} \quad 9\overline{)72} \quad 3\overline{)18}$$

Go on to the next page

"TIMES TABLES THE FUN WAY"
WORKBOOK Lesson 4
WORKSHEET # 2

46	16	56	58	36
x6	x8	x6	x6	x3

64	62	18	27	88
x6	x6	x7	x7	x5

$6\overline{)36}$ $3\overline{)12}$ $8\overline{)48}$ $9\overline{)81}$ $8\overline{)56}$

$3\overline{)18}$ $7\overline{)49}$ $7\overline{)56}$ $5\overline{)40}$ $9\overline{)72}$

$9\overline{)63}$ $9\overline{)54}$ $9\overline{)27}$ $5\overline{)25}$ $6\overline{)18}$

26	28	65	65	33
x6	x6	x6	x8	x7

46	27	17	47	97
x6	x4	x8	x7	x5

6⟌30   7⟌28   8⟌56   8⟌40   8⟌72

6⟌18   4⟌12   6⟌48   6⟌36   7⟌56

3⟌21   7⟌49   9⟌36   5⟌45   3⟌18

"TIMES TABLES THE FUN WAY"
WORKBOOK Lesson 4
WORKSHEET # 2b

46	26	58	56	33
x5	x8	x7	x8	x6

66	26	27	55	38
x5	x6	x7	x7	x6

5)‾30‾   7)‾49‾   6)‾48‾   5)‾40‾   7)‾56‾

3)‾18‾   4)‾12‾   8)‾56‾   9)‾81‾   8)‾72‾

7)‾63‾   6)‾54‾   3)‾27‾   5)‾15‾   6)‾18‾

"TIMES TABLES THE FUN WAY"

GUESS THE FACT

Directions:
1. Write the fact in the blanks below.
2. Write the answer to the fact.
3. Color the pictures.

If you want to be a pilot,
you'll need to learn math.

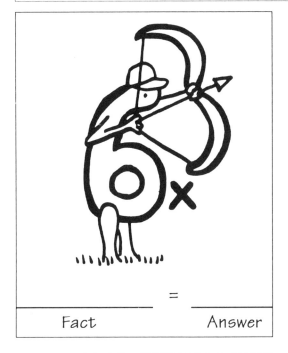

_____ = _____
Fact   Answer

_____ = _____
Fact   Answer

_____ = _____
Fact   Answer

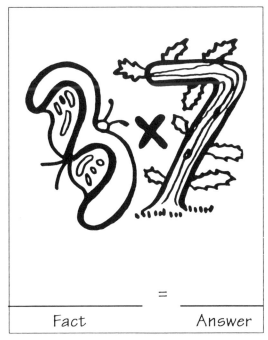

_____ = _____
Fact   Answer

"TIMES TABLES THE FUN WAY"
WORD SEARCH

Directions:
1. The hidden words are listed below.
2. Circle each word you find.
3. Put a check mark by the word once you find it.

I know it's here somewhere.

___Bouncy   ___Cocoon   ___Magic Spell   ___Snowman   ___Cave
___Big Foot   ___Desert   ___High Jump   ___Blind Mice   ___Snail
___Cheer   ___Soldier   ___Hang Glider   ___Mud Pies   ___Cake

S	E	H	A	N	G	G	L	I	D	E	R	O	N
M	R	A	V	H	I	G	H	J	U	M	P	L	A
A	F	L	I	P	D	E	S	E	R	T	A	D	N
R	E	F	O	B	R	I	K	O	Y	E	R	E	C
T	M	A	G	I	C	S	P	E	L	L	Y	A	Y
K	U	M	N	G	A	B	N	E	R	D	S	T	I
I	D	W	E	F	V	E	R	A	Y	O	I	J	S
D	P	O	L	O	E	R	A	S	I	M	C	E	O
S	I	D	B	O	U	N	C	Y	O	L	A	C	R
D	E	L	T	T	A	O	S	H	L	O	K	A	O
O	S	E	A	R	C	W	A	Z	E	X	E	K	O
A	N	S	N	O	W	M	A	N	B	E	S	I	T
C	A	N	O	T	O	R	I	F	C	A	R	E	I
A	D	N	O	B	L	I	N	D	M	I	C	E	P

"TIMES TABLES THE FUN WAY"
WORKBOOK Lesson 5

Answer these facts:

3	7	6	8	7	4	4	3
x 7	x 4	x 8	x 8	x 7	x 3	x 4	x3

CORRECT
_____
8

What is the story for 3 x 7 ? Draw it or write it.

CORRECT
_____
1

What is the story for 7 x 4 ? Draw it or write it.

CORRECT
_____
1

"TIMES TABLES THE FUN WAY"
WORKBOOK Lesson 5
NINES TIMED PRACTICE

# correct:	% score:
$\overline{20}$	

NAME_____DATE_____TIME_____

9 x 5	8 x 9	4 x 9	9 x 9
3 x 9	6 x9	9 x 7	9 x 1
2 x9	9 x 4	5 x 9	9 x 2
9 x 8	9 x 6	7 x 9	9 x 3
7 x7	4 x 3	8 x 8	6 x 4

"TIMES TABLES THE FUN WAY"
WORKBOOK Lesson 5
WORKSHEET # 3

47 x6	56 x3	54 x3	55 x5	46 x4
28 x7	66 x6	58 x5	44 x4	64 x5
73 x5	36 x4	36 x3	76 x5	82 x5

$9\overline{)63}$ $9\overline{)54}$ $9\overline{)27}$ $5\overline{)25}$ $3\overline{)18}$

$21 \div 3 =$ $\qquad$ $25 \div 5 =$

$56 \div 7 =$ $\qquad$ $30 \div 6 =$

"TIMES TABLES THE FUN WAY"
WORKBOOK Lesson 5
WORKSHEET # 3a

56 x7	65 x3	24 x8	95 x5	64 x6
26 x6	67 x6	85 x5	24 x4	55 x5
37 x5	63 x4	37 x3	56 x7	28 x4

$7\overline{)56}$ $6\overline{)42}$ $3\overline{)27}$ $5\overline{)30}$ $6\overline{)18}$

$21 \div 3 =$ $30 \div 5 =$

$45 \div 9 =$ $42 \div 6 =$

"TIMES TABLES THE FUN WAY"
WORKBOOK Lesson 5

STORY QUIZ Write the answer and one or two words about the story.

ANSWER

1. 7 X 7 =

2. 6 X 7 =

3. 8 X 8 =

4. 6 X 4 =

5. 6 X 6 =

6. 6 X 8 =

7. 3 X 7 =

8. 8 X 7 =

9. 8 X 4 =

10. 7 X 4 =

11. 6 X 3 =

67 x4	36 x5	44 x3	55 x5	46 x6
33 x7	67 x6	58 x5	48 x4	76 x6
78 x4	36 x4	36 x3	76 x4	82 x5

$$8\overline{)32} \qquad 7\overline{)28} \qquad 6\overline{)42} \qquad 3\overline{)21} \qquad 6\overline{)18}$$

$$32 \div 4 = \qquad\qquad 35 \div 5 =$$

$$56 \div 8 = \qquad\qquad 48 \div 6 =$$

"TIMES TABLES THE FUN WAY"
HELP SHAGGY DOG
FIND HER WAY HOME

Step One: Fill in the answers to these facts:

7	4	7	4	6	8	7	6
x 6	x 8	x 3	x 7	x 3	x 6	x 8	x 6
___	___	___	___	___	___	___	___
Clue #1	Clue #2	Clue #3	Clue #4	Clue #5	Clue #6	Clue #7	Clue #8

Step Two: Match the answer for each clue with the picture and write the clue word in the blanks below.

48- pound   18 - bloodhound   42 - Mexico   28 - hospital

56 - landlord   21 - bull dogs   32 - greyhound   36-sweet Home

1. Shaggy is lost in _____.
2. Ride the _____ to get back to Utah.
3. Watch out for_____ along the way.
4. Take time-out to heal at the _____.
5. Talk to the _____ to sniff the way home.
6. Get your room-mate out of the _____.
7. Talk to the _____ about signing a new lease.
8. Great! You helped Shaggy find her home _____.

Page 32 - A

"TIMES TABLES THE FUN WAY"
NUMBER GATE MAZE

Directions:
1. Start at the ☒
2. Trace your way through the maze to find the way out. Gates are marked by a ▨
3. You can crash through the gate only if you put the right answer down for the fact.

Don't be stuck in a maze. Learn those times tables!

☒

3 x 7 =

6 x 6 =

8 x 7 =

6 x 8 =

6 x 4 =

8 x 4 =

7 x 6 =

7 x 4 =

6 x 3 =

You made it!

"TIMES TABLES THE FUN WAY"
WORKBOOK Lesson 6

QUIZ #5 NAME_____DATE_____

Answer these facts:

8	7	3	7	6	6	8	6
x 4	x 6	x 7	x 4	x 8	x 3	x 7	x4

CORRECT
___
8

What is the story for 7 x 6 ? Draw it or write it.

CORRECT
___
1

What is the story for 8 x 4 ? Draw it or write it.

CORRECT
___
1

"TIMES TABLES THE FUN WAY"
WORKBOOK Lesson 6
FIVES TIMED PRACTICE

# correct:	% score:
—— 20	

```
   1          5          9          5
 x 5        x 5        x 5        x 4


   6          4          8          5
 x 5        x 5        x 5        x 3


   7          3          5          5
 x 5        x 5        x 2        x 8


   5          5          5          5
 x 1        x 9        x 6        x 7


   3          4          8          7
 x 3        x 3        x 8        x 7
```

"TIMES TABLES THE FUN WAY"
WORKBOOK Lesson 6
WORKSHEET # 4

43	55	77	88	99
x8	x7	x6	x4	x3

77	33	46	36	46
x4	x7	x8	x3	x6

88	78	66	57	43
x7	x8	x4	x7	x3

$36 \div 6 =$ $\qquad$ $35 \div 5 =$

$81 \div 9 =$ $\qquad$ $48 \div 6 =$

38	77	67	98	33
x4	x5	x6	x4	x9

74	23	64	66	47
x6	x7	x8	x3	x6

82	39	72	58	53
x8	x8	x4	x7	x3

$63 \div 9 =$ $35 \div 7 =$

$72 \div 9 =$ $48 \div 8 =$

"TIMES TABLES THE FUN WAY"
WORKBOOK Lesson 6
WORKSHEET # 4b

84	57	87	89	49
x3	x7	x6	x4	x3

47	38	69	96	46
x4	x3	x8	x3	x5

92	87	65	54	49
x7	x8	x4	x7	x3

$48 \div 6 =$ $\qquad$ $32 \div 8 =$

$72 \div 8 =$ $\qquad$ $56 \div 7 =$

"TIMES TABLES THE FUN WAY"
WORKBOOK Lesson 6
HOMEWORK TEST

Name:

Date:

8 x8	7 x7	6 x4	8 x7	6 x6	4 x3	3 x3	4 x4	1 x1
2 x1	3 x2	2 x4	2 x2	2 x9	5 x2	1 x5	4 x1	1 x3
1 x7	2 x6	2 x8	7 x2	6 x1	9 x1	2 x9	5 x3	5 x5
5 x7	5 x9	8 x5	5 x6	5 x4	6 x3	6 x8	3 x7	7 x4
9 x4	6 x9	9 x7	9 x9	9 x8	7 x6	8 x4	8 x3	1 x8

After you have taken this test and graded it, tear this sheet out and learn the missed facts for your homework assignment. Next session, you will have a quiz on the ones you missed.

"TIMES TABLES THE FUN WAY"
WORKBOOK Lesson 6

HOW MANY FACTS DO YOU KNOW?

HOMEWORK QUIZ

QUIZ SCORE

# correct:	% score:

 # MISSED ON HOMEWORK TEST

_____(previous pg.)

Name:

Date:

Write down the facts you missed on the Homework Test here.
Don't put the answer, just the fact. Put one in each box:

Next session you will take this quiz and
if you get 100%, you'll get a stamp.

"TIMES TABLES THE FUN WAY"
GUESS THE FACT

I'm so happy. I know all my times tables!

Directions:
1. Write the fact in the blanks below.
2. Write the answer to the fact.
3. Color the pictures.

____ = ____
Fact   Answer

____ = ____
Fact   Answer

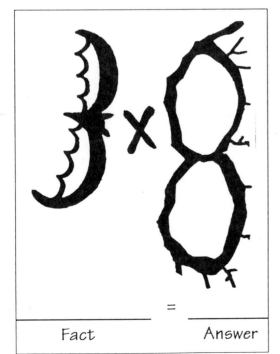

____ = ____
Fact   Answer

In this box, write the fact and answer next to the clue word.

Soldiers ___ x ___ = _____
Thirsty Sixes ___x___ = _____
Birthday Cake ___x___ = _____
Fireman ___x___ = _____
Hunter___x___ = _____
Snail___x___ = _____
Blind Mice___ x ___ = _____
Cheer___x___ = _____
Hang Glider ___x___ = _____
High Jump ___x___ = _____
Butterfly ___x___ = _____
Bat ___x___ = _____
Mud Pies ___x___ = _____
Snowman ___x___ = _____
Bouncy ___x___ = _____

"TIMES TABLES THE FUN WAY"
FILL IN THE BLANK

6 x 6

Isn't that the one about the thirsty sixes?

Directions:
1. Fill in the blanks with the words listed . Put a check mark by the word once you've used it.
2. Write the answer to the fact.

1. Two 6's were travelling in the _____ when they found _____.
   (6 x 6 = _____)
2. The two 7's are _____ who watch for people who _____.
   (7 x 7 = _____)
3. Three blind _____ were born with _____ tails each.
   (3 x 3 = _____)
4. The 7 _____ on his 8 _____.
   (7 x 8 = _____)
5. The wicked _____ put a spell on the _____.
   (6 x 8 = _____)
6. Bart saved his _____ to buy a _____.
   (4 x 4 = _____)
7. The snail made a _____ and became a _____.
   (6 x 4 = _____)
8. Farmer John saved the _____, but the _____ burned.
   (7 x 4 = _____)
9. The 8-pigs took a _____ because they were _____.
   (8 x 4 = _____)
10. The two snowmen got _____ so they built a _____.
    (8 x 8 = _____)
11. One, two, three, _____, I like _____ let's do some more.
    (3 x 4 = _____)
12. The bat was _____ because he saw the King of _____.
    (3 x 8 = _____)
13. The _____ landed in front of _____.
    (6 x 3 = _____)
14. _____ like to hang out at the _____ tree.
    (3 x 7 = _____)
15. The 7 is _____ and the 6 is very _____.
    (7 x 6 = _____)

___Big Foot
___Butterflies
___four
___seven
___bounces
___scared
___witch
___mice
___soldiers
___water
___chubby
___litter
___money
___bath
___desert
___skinny
___trampoline
___swan
___snakes
___barn
___wish
___animals
___arrow
___three
___fire
___dirty
___cold
___math
___4 x 4
___castle

QUIZ #6 NAME_____DATE_____

Answer these facts:

8	7	8	8	6	9	8	6
x 3	x 6	x 4	x 8	x 8	x 9	x 7	x 4

CORRECT
____
8

What is the story for 8 x 3? Draw it or write it.

CORRECT
____
1

"TIMES TABLES THE FUN WAY"
WORKBOOK Lesson 7
NINES TIMED PRACTICE

# correct:	% score:
$\dfrac{}{20}$	

NAME_____DATE_____TIME_____

9 x 5	8 x 9	4 x 9	9 x 9
3 x 9	6 x9	9 x 7	9 x 1
2 x9	9 x 4	5 x 9	9 x 2
9 x 8	9 x 6	7 x 9	9 x 3
7 x7	4 x 3	8 x 8	6 x 4

"TIMES TABLES THE FUN WAY"
WORKBOOK Lesson 7
Worksheet # 5

78 88 x4 x3	Name two numbers that will give you this answer when they are multiplied together. 20 81 36 25
Division: 5 ⟌ 25 6 ⟌ 36 4 ⟌ 24 3 ⟌ 18	What is your favorite story? Why?
Draw a picture of your favorite story here:	$7 \times 8 =$ $8 \times 8 =$ $7 \times 7 =$ $7 \times 6 =$
Story Problem: There was a boy named Pete. He had two mice for pets. Each mouse ate 25 cents worth of food each day. 1. How much did it cost Pete to feed his mice for a week?	2. How much did it cost to feed the mice for a month? (4 weeks)

"TIMES TABLES THE FUN WAY"
WORKBOOK Lesson 8
CROSSWORD PUZZLE

ACROSS		DOWN	
1. 4 x 4 = _____	7. 6 x 8 Story	1. 7 x 7 Story	7. 7 x 8 Story
2. 6 x 7 Story	8. 6 x 3 Story	2. Camping House	8. 6 x 6 Story
3. 8 x 4 Story	9. A boy or	3. 8 x 3 Story	9. 7 x 5 = _____
_____pigs.	girl's name.	4. Grown up guy	10. 8 x 8 Story
4. 6 x 4 Story	10. 3 x 7 Story	5. When you're fin-	11. 2 x 5 = _____
5. 4 x 2 = _____	11. 3 x 3 Story	ished, you're _____	12. 3 x 4 Story
6. 7 x 4 Story	12. Opposite of no	6. Stop and _____	

"TIMES TABLES THE FUN WAY"
WORKBOOK Lesson 8
POST-TEST

# correct:	% score:
___ 48	

NAME_____DATE_____TIME_____

2 x1	3 x2	4 x4	3 x6	7 x8	8 x9	1 x3	2 x4
5 x3	6 x4	8 x8	9 x4	1 x1	2 x2	5 x4	6 x6
9 x3	6 x9	7 x4	2 x9	4 x1	5 x2	5 x5	8 x6
9 x7	5 x7	9 x1	1 x5	0 x7	3 x7	6 x1	6 x7
8 x5	9 x9	3 x8	5 x9	4 x3	8 x1	7 x2	3 x0
7 x7	1 x7	2 x6	8 x4	5 x6	0 x9	3 x3	2 x8

"TIMES TABLES THE FUN WAY"
STAMP AND SCORE SUMMARY SHEET

Lesson 1

Pre-test
Score: _____
Time: _____

1's & 2's Practice
Score: _____
Time: _____

Stamps:

100% 1's & 2's		

Lesson 2

Quiz # 1
Facts:_____ Stories:_____

Fives Practice
Score: _____
Time: _____

Stamps:

100% Fives	Workbk. page 11	Workbk. page 12
Workbk. page 13	Workbk. page 14	Won Game
Won Game		

Lesson 3

Quiz # 2
Facts:_____ Stories:_____

1's & 2's Practice
Score: _____
Time: _____

Nines Practice
Score: _____
Time: _____

Stamps:

100% 1's & 2's	Improved Score 1's & 2"s	Improved Time 1's & 2's
100% Nines	Workbk. page 17	Workbk. page 18
Won Game		

Lesson 4

Quiz # 3
Facts:_____ Stories:_____

Fives Practice
Score: _____
Time: _____

Stamps:

100% Fives	Improved Score Fives	Improved Time Fives
Workbk. page 22 page 23	Workbk. page 24 page 25	Workbk. page 26
Won Game	Won Game	Won Game

Lesson 5

Quiz # 4
Facts:_____ Stories:_____

Nines Practice
Score: _____
Time: _____

Story Quiz
Facts: _____ Stories:_____

Stamps:

100% Nines	Improved Score Nines	Improved Time Nines
Workbk. page 29	Workbk. page 30	Workbk. page 32
100% Story Quiz Stories	100% Story Quiz Facts	Won Game

Lesson 6

Quiz # 5
Facts:_____ Stories:_____

Fives Practice
Score: _____
Time: _____

Stamps:

100% Fives	Improved Score Fives	Improved Time Fives
Workbk. page 35	Workbk. page 36	Workbk. page 37
100% Homework Test	Won Game	Won Game

Lesson 7

Quiz # 6
Facts:_____ Stories:_____

Homework Quiz:
Score: _____

Nines Practice
Score: _____
Time: _____

Stamps:

100% Nines	Improved Score Nines	Improved Time Nines
Workbk. page 42	100% Homework Quiz	Returned Signed HW Test
Won Game	Won Game	

Lesson 8

Post-test
Score: _____
Time: _____

Stamps:

Workbk. page 43	100% Post-Test	Won Game

"TIMES TABLES THE FUN WAY"
Workbook Answer Sheet

Pre-test/ Post-test Page 2, 44

2	6	16	18	56	72	3	8
15	24	64	36	1	4	20	36
27	45	28	18	4	10	25	48
63	34	9	5	0	21	6	42
40	81	24	45	12	8	14	0
49	7	12	32	30	0	9	16

Ones and Twos Page 3, 16

1	2	3	6
8	9	7	5
4	4	10	12
18	16	14	8
6	9	12	497

Guess the Fact Page 3A

7 x 7 = 49	3 x 3 = 9
8 x 8 = 64	3 x 4 = 12

Pirate Finds Treasure Page 3B

12	8	14	16	10	6	18	3

1. Africa	2. Dolphin
3. Sharks	4. Donkey
5. Wise Owl	6. Mt. Top
7. Treas. Map	8. Treasure

Quiz # 1 Page 4

9	49	12	64
Snowman		Mice	
Soldiers		Cheer	

Fives Timed Practice Page 5, 21, 34

5	25	45	20
30	20	40	15
35	15	10	40
5	45	30	35
9	12	64	49

Learning Double Digit Page 7

6				
126				
35	46	126	68	48
29	93	96	164	86

Double Digits-Regrouping Page 9

252				
50	72	56	154	212

Double Digits-cont. Page 10

372	656	126	189	275
0	0	0	0	0,etc

Tens, Eleven and Twelves Page 11

22	30	36	20	11	24	50
60	70	72	77	99	90	48
12	80	88	10	66	108	40
96	33	44	60	55	50	24

Worksheet #1 Page 12

36	44	42	99	119
144	34	154	132	96
72	72	99	108	176
189	224	84	60	48

Worksheet #1a Page 13

75	180	129	156	399
224	94	110	128	165
156	132	189	396	616
196	216	504	260	128

Worksheet #1b Page 14

156	220	156	260	189
144	116	330	64	175
130	132	69	366	616
539	224	574	360	96

Guess The Fact Page 14A

6 x 4 = 24	6 x 6 = 36
4 x 4 = 16	7 x 8 = 56

Color Dufus Page 14B

10 x 1 = 10	2 x 5 = 10	1 x 5 = 5
7 x 5 = 3	3 x 5 = 15	9 x 5 = 45
4 x 5 = 20	5 x 3 = 15	15 x 3 = 45
8 x 5 = 40	6 x 5 = 30	5 x 6 = 30

Quiz # 2 Page 15

24	16	56	36	64	49	9	12
Snails				Bouncy Guy			
Driver's License				Thirsty Sixes			

Learning Nines Page 17

2	6	7	5	3	4	8	1
7	5	4	3	0	2	1	6

Learning Nines Page Two Page 18

63	54	27	18	9	72	63
72	9	54	81	63	45	36
27	18	27	18	45	54	81
63	72	18	36	45	81	72

Nines Timed Practice Page 19, 28, 41

45	72	36	81
27	54	63	9
18	36	45	18
72	54	63	27
49	12	64	24

Color The Wizard Page 19A

9 x 9 = 81	9 x 4 = 36	9 x 5 = 45
9 x 2 = 18	27 x 3 = 81	3 x 3 = 9
9 x 1 = 9	12 x 3 = 36	15 x 3 = 45
9 x 7 = 63	9 x 5 = 45	9 x 8 = 72
6 x 9 = 54		

Connect the Dots Page 19B

Parrot

Quiz # 3 Page 20

18	48	25	30	81	63
40	36				

Hunter   Birthday Cake

Learning Division Page 22

5	4	4	6	5	7
9	8				

Learning Division Page 2 Page 23

4	4	8	8	6

Worksheet #2 Page 24

276	128	336	348	98
384	372	126	189	440
6	4	6	9	7
6	7	8	8	8
7	6	3	5	3

Worksheet #2a Page 25

156	168	390	520	231
276	108	136	329	486
5	4	7	5	9
3	3	8	6	8
7	7	4	9	6

Worksheet #2b Page 26

230	208	406	448	198
330	156	189	385	228
6	7	8	8	8
6	3	7	9	9
9	9	9	3	3

Guess The Fact Page 26A

6 x 3 = 18	6 x 4 = 28
6 x 8 = 48	3 x 7 = 21

Word Search Page 26B

S	E	H	A	N	G	G	L	I	D	E	R	O	N
M	R	A	V	H	I	G	H	J	U	M	P	L	A
A	F	L	I	P	D	E	S	E	R	T	A	D	N
R	E	F	O	B	R	I	K	O	Y	E	R	E	C
T	M	A	G	I	C	S	P	E	L	L	Y	A	Y
K	U	M	N	G	A	B	N	E	R	D	S	T	I
I	D	W	E	F	V	E	R	A	Y	O	I	J	S
D	P	O	L	O	E	R	A	S	I	M	C	E	O
S	I	D	B	O	U	N	C	Y	O	L	A	C	R
D	E	L	T	T	A	O	S	H	L	O	K	A	O
O	S	E	A	R	C	W	A	Z	E	X	E	K	O
A	N	S	N	O	W	M	A	N	B	E	S	I	T
C	A	N	O	T	O	R	I	F	C	A	R	E	I
A	D	N	O	B	L	I	N	D	M	I	C	E	P

Quiz # 4 Page 27

21	28	48	64	49	12
16	9				

Butterflies   Burning Barn

Worksheet #3 Page 29

282	168	162	275	184
196	396	290	176	320
365	144	98	380	410
7	6	3	5	6
7	5			
8	5			

Worksheet #3a Page 30

392	195	192	475	384
156	402	425	96	275
185	252	111	392	112
8	7	9	6	3
7	6			
5	7			

Story Quiz Page 31

49	Soldiers	42	High Jump
64	Snowman	24	Snail
36	Thirsty Sixes	48	Birthday Cake
21	Cocoon	56	Trampoline
32	Dirty Pigs	28	Too Late
18	Hunter		

Worksheet #3b Page 32

268	180	132	275	276
231	402	290	192	456
312	144	98	304	410
4	4	7	7	3
8	7			
7	8			

Help Shaggy Dog Page 32A

42	32	21	28	18	48	56	36

1. Mexico   2. greyhound
3. bull dogs   4. hospital
5. bloodhound   6. pound
7. landlord   8. sweet home

Number Gate Maze Page 32B

3 x 7 = 21	6 x 6 = 36	8 x 7 = 56
6 x 8 = 48	6 x 4 = 24	7 x 6 = 42
8 x 4 = 32	7 x 4 = 28	6 x 3 = 18

Quiz # 5 Page 33

32	42	21	28	48	18
56	24				

High Jump   Dirty Pigs

Worksheet #4 Page 35

344	385	442	352	297
308	231	368	108	276
616	624	264	399	129
6	7	9	8	

Worksheet #4a Page 36

152	385	462	352	297
444	161	512	198	262
656	312	288	406	159
7	5	8	6	

Worksheet #4b Page 37

252	399	522	356	147
188	114	562	288	230
644	696	260	378	147
8	4	9	8	

Homework Test Page 38

64	49	24	56	36	12	9	16	1
2	6	8	4	18	10	5	4	3
7	12	16	14	6	9	18	15	25
35	45	40	30	20	18	48	21	28
36	54	63	81	72	42	32	24	8

Guess the Fact Page 39A

$8 \times 4 = 32$	$7 \times 6 = 42$	
$3 \times 8 = 24$		
$7 \times 7 = 49$	$6 \times 6 = 36$	$6 \times 8 = 48$
$7 \times 4 = 28$	$6 \times 3 = 18$	$6 \times 4 = 24$
$3 \times 3 = 9$	$3 \times 4 = 12$	$7 \times 6 = 42$
$3 \times 7 = 21$	$3 \times 8 = 24$	$4 \times 8 = 32$
$8 \times 8 = 64$	$7 \times 8 = 56$	

Fill in the Blank Page 39-B

1. desert	water	$6 \times 6 = 36$
2. soldiers	litter	$7 \times 7 = 49$
3. mice	three	$3 \times 3 = 9$
4. bounces	trampoline	$7 \times 8 = 56$
5. witch	castle	$6 \times 8 = 48$
6. money	4 x 4	$4 \times 4 = 16$
7. wish	swan	$6 \times 4 = 24$
8. animals	barn	$7 \times 4 = 28$
9. bath	dirty	$8 \times 4 = 32$
10. cold	fire	$8 \times 8 = 64$
11. four	math	$3 \times 4 = 12$
12. scared	snakes	$3 \times 8 = 24$
13. arrow	Big Foot	$6 \times 3 = 18$
14. butterflies	seven	$7 \times 3 = 21$
15. skinny	chubby	$6 \times 7 = 42$

Quiz # 6 Page 40

24	42	32	64	48	81	56	24

Bat in the cave

Worksheet #5 Page 42

312		264	
4 x 5		9 x 9	
4 x 9		5 x 9	
5	6	6	6
56	64	49	42

1. $3.50   2. $14.00

Crossword Puzzle Page 43

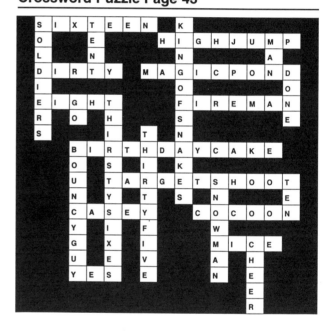